Dubai 2075

01, Volume 1

Rajuan Ahmed

Published by Headroom Publishers, 2026.

While every precaution has been taken in the preparation of this book, the publisher assumes no responsibility for errors or omissions, or for damages resulting from the use of the information contained herein.

DUBAI 2075

First edition. July 30, 2026.

ISBN: 979-8235132757

Written by Rajuan Ahmed.

Also by Rajuan Ahmed

01
Beyond the Naf
Dubai 2075
Exiled Dreams
Quantum Wizards
In the Land of Mirages
Love of the Forest Bird
The Days of Gaza
Geneva Camp
We Are All Mad
The Shadow of War

Standalone
Death Break

To every soul who chooses hope over fear, compassion over comfort, and life over despair. May this story honor those who plant the seeds of tomorrow, even as the world around them burns.

Dubai 2075

By Rajuan Ahmed

DEDICATION

To every soul who chooses hope over fear, compassion over comfort, and life over despair. May this story honor those who plant the seeds of tomorrow, even as the world around them burns.

Summary

By 2075, Dubai has transformed into an environmental nightmare. Dr. Alia Al-Qasemi, a 55-year-old climate scientist, witnesses the city's daily struggle against extreme heat and humidity that makes human survival a relentless battle. The wet-bulb temperature regularly exceeds 36°C, a threshold fatal to the human body.

Alia, born to an Emirati father and a Bangladeshi mother, works at the Dubai Institute of Climate Research. Although her son, Hamza, lives in London and repeatedly begs her to relocate, she remains deeply committed to her city and her research. Meanwhile, as the Persian Gulf reaches unprecedented temperatures, desalination plants collapse, crippling the water supply. Frequent power outages trigger a vicious cycle that threatens the city's remaining infrastructure.

The city's migrant labor force—primarily workers from Bangladesh, Pakistan, India, Nepal, and the Philippines—bears the brunt of the crisis. Over 2,500 laborers died from heatstroke the previous year alone. Alia personally saves one such worker, Mohammad Idris from Comilla, Bangladesh, after finding him on the verge of death from extreme heat on a public street.

When Cyclone Harmattan strikes Dubai with unprecedented fury, it exposes the city's extreme vulnerability. Fueled by warming ocean waters, the devastating storm collapses critical infrastructure and utterly destroys the migrant laborer camps. In the grim aftermath, Alia realizes that while the wealthy elite have fled, the poor are left behind to pick up the pieces.

In response, Alia launches "Green Roots," a community garden project designed to provide displaced workers with food, training, and purpose. Idris steps up as its supervisor, finding personal renewal through working with the earth. The initiative quickly spreads across the city, teaching practical skills and sustainable agriculture techniques.

However, years of breathing Dubai's toxic air cause Alia to develop severe pulmonary fibrosis. Despite her deteriorating health, her doctors' warnings, and her son's heartfelt pleas, she refuses to abandon her post. Instead, she pours her thoughts into a notebook, a testament and warning to future generations about how human hubris attempted to conquer nature. She documents the city's fatal illusion: believing that money and technology could ever replace environmental balance.

In November 2075, Alia passes away, but her vision lives on. Hamza dedicates his life to continuing her work, expanding Green Roots across the UAE and eventually around the world. Idris's daughters receive an education and carry her legacy forward, while Alia's published notebook, titled The Last Letter of the Century, sparks a global environmental movement.

The story concludes at Alia Garden, a flourishing sanctuary in Deira that serves as a living testament that hope can bloom even in the desert. Ultimately, the narrative reinforces a powerful lesson: love, rather than technology, is humanity's greatest key to survival, and small, compassionate actions can seed lasting global change.

The city had felt like a suffocating furnace since dawn.

Standing by the window on the forty-fifth floor, Alia Al-Qasemi looked down. The moment her palm touched the glass, she could feel how desperately the outside heat was trying to breach the interior; the pane felt like a searing iron skillet. Down on Sheikh Zayed Road, vehicles glided along like bubbles trapped on a track entirely encased in glass, moving on hushed electric engines. Yet, an odd exhaustion lined the faces of the people sitting inside them, as if they were not traveling at all, but merely drifting. Sunlight slammed against the glass towers with a blinding, violent glare. There was a time when this city felt like a dream; now, it felt like a mirage a strange, artificial illusion whose core was slowly rotting into emptiness.

Around three o'clock the previous night, the air conditioner's compressor finally gave out. First, its hum shifted a muffled, groaning sound and then it went completely dead. Alia glanced at the clock: 9:30 AM. Inside the apartment, the temperature had already climbed past 43°C. She hadn't yet dared to check the conditions outside, but yesterday afternoon she had looked toward Al Maktoum International Airport and watched intense waves of heat shimmering violently over the runways. It felt as though the entire city had been shoved inside a massive tandoor oven, with tongues of fire licking up from below. A few birds were flying or trying to but they kept sinking into the thermal currents, as if an invisible hand were weighing down their wings.

Her gaze shifted to the digital thermometer on the table. 45°C, with humidity at 85%. The algorithm-driven screen mounted on the wall delivered the verdict: the wet-bulb temperature the combined index of humidity and heat stood at 36.2°C. No one understood better than Alia that this number was a human death warrant. She was a climate scientist, the chief researcher at the Dubai Institute of Climate Research. For the past fifteen years, she had gone to sleep and woken up to these very numbers. Ever since her husband passed away, these cruel, silent, yet honest digits had become her only true companions.

Her phone began to vibrate on the bedside table. Hamza's name flashed across the screen. Her son was video-calling from London. He called at least three times a week, and every single time, he said the exact same thing: "Ammi, please leave. Come here."

Alia picked up the phone. She forced a smile for the camera, keeping her voice steady, pretending everything was perfectly fine. "Everything will be alright, baba. The government announced that the cloud-seeding project will be fully operational by this week. The rain will come, and things will return to normal."

"What rain, Ammi? You're the one who said cloud seeding doesn't work anymore. There isn't enough moisture left in the atmosphere for clouds to form. Even if they do, the rain evaporates in the scorching air before it ever hits the ground. Are you denying your own research now?"

Hamza's words struck a sharp, aching chord in Alia's chest. Her son was right. But how could she admit it? To admit it was to accept that this city was dying, and that she was fading away right along with it.

After hanging up, Alia went to the bathroom and turned the tap. Nothing came out. At first, she thought it was just air trapped in the pipes. She waited a few moments, hoping for a delayed trickle. But no, the dry faucet offered only a faint, hollow hiss before falling silent. The water supply had been cut off since the previous night. The news reported severe complications at the Jebel Ali Desalination Plant. The waters of the Persian Gulf had warmed to such an unprecedented degree that the plants were struggling to process the seawater; a massive amount of electricity was required just to cool the water down before filtration could even begin.

Power shortages were now a daily reality the city's rolling blackouts were as meticulously punctual as the hands of a clock. A thick layer of desert dust had settled over the panels at the Mohammed bin Rashid Al Maktoum Solar Park, slashing power generation by nearly 30%. Yet, there was no water to clean the dust. It was a cruel paradox: the lack of water left the panels dirty, the dust reduced the power, and the lack of power meant there was no electricity to run the desalination plants. It was a vicious cycle from which no one had yet figured out an escape.

A notification chimed on Alia's phone a message from the Dubai Municipality, the same text that arrived three or four times a week: "All residents are advised to remain indoors between 10:00 AM and 4:00 PM. The wet-bulb temperature has exceeded critical, life-threatening thresholds."

During the peak of summer, people had grown accustomed to this. Those without urgent business stayed inside watching television in air-conditioned apartments, browsing the internet, or sleeping. But those driven by sheer survival the laborers, construction workers, delivery riders, cleaners, and restaurant cooks had no choice but to go out. For them, the city had become a death trap. Last year alone, more than two and a half thousand laborers died of heatstroke in Dubai. Most of them had come from Bangladesh, Pakistan, India, Nepal, and the Philippines. No one spoke of their deaths; they were never splashed across the front pages of newspapers. Because this was a silent passing dying from heatstroke wasn't a terrorist attack, a building collapse, or a plane crash; it was nature's quiet execution. And you cannot bring charges against nature.

Alia changed and got ready for the office. She put on a light-colored cotton saree as thin and loose-fitting as possible. She had inherited this habit of adapting to the weather from her Bangladeshi mother. Her father was an Emirati businessman, her mother a girl from Sylhet. As a child, she used to spend her summer vacations with her mother in Sylhet. She had seen the mist-covered tea gardens, drenched herself in the monsoons, and listened to the rhythmic dance of rain on corrugated tin roofs. Her mother used to say, "Can you smell the Sylhet rain? You won't find this scent anywhere else in the world." That memory now felt like a dream. Sylhet wasn't the same anymore either shattered by a bizarre, volatile pendulum swing between floods and droughts. Yet, the Bangladesh she had envisioned through her mother's stories remained eternally green in her imagination.

On her way down in the elevator, she ran into her neighbor, Mrs. Rogers. The Australian woman was nearing sixty, her hair long since turned gray. Her husband had been an Emirates airline pilot who died of cancer the previous year. She had spoken of leaving Dubai to return to Melbourne, but she was still here. Alia didn't know exactly why, though she could guess perhaps there were no safe places left on Earth, or perhaps she simply couldn't gather the strength to abandon the memories woven into every corner of this city.

"Any news, Alia? I heard the government is building a new underground city outside the main grid. The entire city will be subterranean, with artificial sunlight and fully controlled temperatures."

Alia offered a soft, skeptical smile. "The plans have been around for a long time. But it will take years for construction to actually begin. Right now, it's just a pilot project. And as you know, there is a vast gulf between Dubai's grand designs and their realization."

"I heard this city will have an artificial sky, artificial air, even artificial rivers everything perfectly regulated! Humanity will no longer be at the mercy of nature."

"Mrs. Rogers, Dubai has always loved to dream. It built artificial islands, artificial ski slopes. Perhaps now it will build an artificial world."

But in her heart, Alia knew the lifespan of this artificial dream was running out.

The moment she stepped outside, a blast of scorching air caught in her throat, cutting off her breath. Alia quickly covered her face with a handkerchief. The air was so heavy with moisture that the fabric became damp within seconds; yet, because of the stifling heat, that dampness offered no comfort it only heightened her anxiety, making breathing even more labored. As she reached to open her car door, she noticed a construction worker sitting by the curb. He wore a blue uniform with the company name printed across the back: Al-Futtaim Construction. Sweat was pouring down the man's face, his eyes were bloodshot as hibiscus petals, and his body was trembling violently. Beside him lay a cracked plastic water bottle completely empty. These were the classic signs of heatstroke. Alia quickly grabbed a fresh bottle of water from her car and rushed over.

"Brother, drink some water."

The man looked up. He was Bangladeshi. Not a face she knew, yet his eyes held a haunting familiarity that eternal exhaustion, that devastating hope that Alia saw on the streets of Dubai every single day. Her heart wrenched. The face felt so painfully close, like a lost relative from her mother's homeland.

"How are you feeling? Drink some water, please. Sip it slowly, don't gulp it all at once."

The man took the bottle with a shaking hand. He emptied half of it in his very first gulp. Then, taking a long, ragged breath, he took another sip. "Sister, save me. I can't work. My head is spinning from the heat and I'm falling over. But the supervisor says if I don't finish the work, they won't pay my wages."

Alia closed her eyes. Who was responsible for the plight of these workers? The contractor? The employer? The government? Or the entire global system a system where the lives of the poor are laid down like bricks to sustain the glittering illusion of petrodollars? She called for an ambulance. The operator responded, "It will take at least two hours. We are backed up with calls. If you can, bring him to the hospital yourself."

Alia made up her mind; she would drive the worker to the hospital herself. His name was Mohammad Idris, a young man from Comilla. His wife and two daughters lived back in their village the eldest, Nadia, was eleven; the youngest, Nishat, was seven. Idris had come to Dubai four years ago. He had hoped to save some money, return to his village, educate his daughters, and open a small shop. But he hadn't been able to save anything yet. The company paid wages erratically sometimes on the fifteenth of the month, sometimes on the fifth of the following month, and occasionally skipping a month entirely. The contractor had confiscated his passport. Idris was trapped. Not on paper slavery was strictly outlawed by the Emirati Constitution but in reality, he was bound by the invisible chains of modern servitude.

It took nearly two hours to get Idris admitted. The emergency ward at Rashid Hospital was bursting at the seams every bed was occupied, stretchers lined the corridors, and groaning patients writhed in every corner. Finally, after an exhausting ordeal, they managed to secure a bed at Maktoum Hospital. By the time Alia reached her office, it was one o'clock in the afternoon.

Stepping into the laboratory, she found her colleagues huddled together, their faces clouded with anxiety. Rashed, a junior researcher, stepped forward. "Dr. Alia, the satellite data arrived this morning. You won't believe it the water temperature in the Persian Gulf rose by 0.5°C in just the last week. It's a record." His voice carried less excitement and far more terror.

Alia sank into her chair. The Persian Gulf was already one of the warmest bodies of water on the planet. Last year, it had hit a record peak of 38°C; now, it was poised to shatter that. Alia knew exactly what this meant an amplification of tropical cyclones, the complete, near-total annihilation of coral reefs, the collapse of marine breeding cycles, and a catastrophic surge in humidity. And as humidity climbed, the wet-bulb temperature would follow; when that happened, simply stepping onto the street would mean stepping into the arms of death. It was a complex equation where every calculation pointed toward mortality.

Alia logged into her terminal and began analyzing the data. Every graph, every digit felt like a silent scream the Earth warning them that time had run out, that they needed to wake up immediately. The temperature curve flashing on the screen looked like someone leaping off a precipice slowly at first, then gaining momentum, before plunging straight down.

But who was listening to the scream? In this city, everyone knew the world was ending; yet, they simply shopped more, partied harder, and dined in luxury hotels. Dubai's mega-malls the Dubai Mall, the Mall of the Emirates, the Marina Mall still thrummed with dense, bustling crowds. Stepping into those air-conditioned, enchanted sanctuaries, it was impossible to tell that the world outside was burning to ashes. This disconnect was perhaps Dubai's profoundest tragedy a city that knew it was dying, yet pretended it was immortal. Lying on its deathbed, it still painted its lips, looked in the mirror, and smiled.

Returning home that evening, Alia opened the notebook resting on her desk. On the very first page, she had written in English: "Last Letter of the Century." It was an ordinary notebook no leather binding, just a plastic cover protecting lined white paper. Yet, this notebook had become her only sanctuary. In it, she wrote letters. Not to her son Hamza, but to some unknown future reader someone who might read it at the end of this century, or perhaps at the start of the next. Alia believed that one day, someone would read her words. And on that day, humanity would understand how this world was undone, what the people living through it were thinking, and how they clung to the very last remnants of hope.

She picked up her pen. It was an old pen, its plastic casing hairline-fractured, but it still wrote smoothly. Her husband had given it to her many years ago for her birthday. He had told her, "Whatever you write, write it with this pen. This is your voice." Alia began to write:

Dear Reader of the Future,

By the time you read this letter, a city named Dubai may no longer exist. Where the Burj Khalifa once pierced the heavens, there may be nothing now but shifting sand dunes, ruins, and the briny wash of the sea. Where Palm Jumeirah once stood as the ultimate monument to luxury, there may be nothing left but crabs and jellyfish. Where millions once flocked to chase their dreams, there may remain only silence and the harsh desert wind.

I do not know who you are. You might be a historian researching the archaeology of this city, or perhaps a schoolchild learning stories of an old world. But I want you to take away not just the history, but the lesson. Understand why this city fell, who was to blame, and how you might build a new world out of our failures.

I am Alia Al-Qasemi. I am fifty-five years old, a climate scientist by profession. I was born in Dubai to a Bangladeshi mother and an Emirati father. I love this city. But I also know that the dream this city is anchored upon is fundamentally unsustainable. Or rather, the lie this city is built on is unraveling. Every day, every hour, nature exposes that lie.

Outside, the sun has set, but the temperature remains stuck at 42°C. The air conditioning is off to conserve power, it won't be turned on until 10:00 PM. I am writing by the window. The ceiling fan is spinning, though it offers nothing more than theoretical comfort; the air it moves is hot, humid, and thick with dust. Suspended in this dust are particles of cement from construction sites, the microscopic wear of car tires, and the fine sand of the desert. We breathe this dust into our lungs with every breath. A study conducted last year on respiratory illnesses revealed that 63% of adults living in Dubai show symptoms of pulmonary fibrosis the hardening of lung tissue. Asthma rates among children have doubled. We are slowly suffocating to death, and no one utters a sound.

Alia paused. Her hand was trembling. The intense humidity made it difficult for the ink to set on the page, the characters bleeding slightly into the paper. Yet she kept writing; she couldn't stop. To stop now felt like a crime.

The next morning, she was jolted awake by the harsh ringing of her phone. When she answered, Dr. Rashed's frantic voice came rushing through the line.

"Dr. Alia, you need to get to the lab immediately. New data just came in. The projections we made about the Gulf Stream slowdown are happening much faster in reality! We thought it wouldn't get this bad until 2080, but the satellite data shows the situation is already critical."

Alia dressed quickly and rushed to the lab. The streets were unusually quiet today most people had stayed home, as the municipality had issued yet another emergency warning. Rashed pointed to a graph on a monitor. As ocean temperatures rose, the Atlantic Meridional Overturning Circulation AMOC was decelerating. This current regulated the weather of the North Atlantic, and its disruption triggered a domino effect across the global climate. But its regional repercussions on landlocked waters like the Persian Gulf were proving catastrophic the slowing current was locking intense humidity into the region, suppressing rainfall, and driving temperatures to anomalous extremes. It felt like an invisible monster exhaling hot breath down their necks.

"If this trajectory continues, agricultural production across the entire Middle East will become impossible within the next ten years," Rashed warned. "Not just in Dubai; the coastal regions of Kuwait, Qatar, Bahrain, and Saudi Arabia will face devastating conditions. People will become entirely dependent on imports for food, and as transportation costs skyrocket, food prices will become prohibitive."

Alia sank into her chair. She remembered when she was pursuing her PhD at the Massachusetts Institute of Technology (MIT) twenty years ago. Her supervisor, Professor Richardson, had told her, "Your research topic is fascinating, Alia. But people won't want to think about it. To talk about climate change is to frighten people. And if you frighten them, they won't vote for you. They will always vote for the politician who smiles and says, 'Everything is fine, there's nothing to worry about.'"

That prophecy had been fulfilled to the letter. Humanity had been afraid, but they chose denial. And now, that fear had manifested into physical reality. The escape routes were closing; no place left on Earth was truly safe anymore.

In the afternoon, Alia drove out to Deira, the older quarter of the city. Here lay the gold souk, the spice shops, and older residential buildings structures whose walls bore cracks, yet inside, life still pulsed vibrantly. There was a time when this district was the beating heart of Dubai; now, it felt like a living museum, the final witness clinging to the memories of a bygone era. A few Bangladeshi, Pakistani, and Indian shops still endured here small eateries from which the rich aromas of biryani and chicken polao drifted into the heat.

Alia sat down in one of the restaurants and ordered a biryani. The owner, Haji Rashid, was a Pakistani gentleman in his late sixties with a long, flowing white beard; his eyes showed deep weariness, but his lips held an unyielding smile. He wore a white shalwar-kameez stained with curry the badge of a busy kitchen. Seeing Alia, he smiled warmly.

"Any special news, Madam? One look at your face tells me something heavy is weighing on your mind. Sit, let me get you some hot biryani."

Alia smiled back. "Everyone is worried, Haji Sahab. How is business holding up in this heat?"

"We are just barely surviving, to be honest. The tourists don't come anymore. It used to be that Japanese, Chinese, and European tourists arrived in droves. They came to see the spice souk, buy jewelry from the gold market, and sit right here to eat my biryani. Now, no one wants to come. They say the heat in Dubai takes the life out of you."

Haji Rashid sighed, adjusting his apron. "Even in winter, the temperature doesn't drop below thirty anymore. December and January used to bring foreigners chasing the mild winter sun; now, winter has ceased to exist. I'm thinking of returning to Pakistan. But the situation isn't any better there. People are dying from the heat in Lahore fifty people passed away in a single day last year. Sindh is perpetually flooded, and Karachi has no water. Where is left to go?"

That was the question on everyone's lips: Where is left to go? Most regions of the world had become either too hot, too wet, or completely arid. The temperate zones had limited space and had adopted draconian immigration policies. In Europe, far-right factions had swept into power, sealing borders with barbed wire, concrete walls, and standing armies. The United States had enacted a "fortress policy" where every migrant was treated as a hostile suspect. Countries like Canada and New Zealand had reduced their immigration quotas so drastically that entry was impossible without winning a literal lottery. Humanity was trapped in their own nations, their own cities like survivors clinging to the last floating planks of a shipwreck. And all around them lay water, nothing but boundless, salty water.

Leaving the restaurant, Alia walked down Khalid Bin Waleed Road. By the sidewalk, she noticed a group of laborers lined up to buy water. A tiny shop, owned by an elderly Punjabi man, was doing a brisk business; he pulled bottles from the shelves as cash clinked into the register. A single bottle of water now cost five dirhams; just a year ago, it had cost only one. The price of water had skyrocketed due to the global energy crisis. Running the desalination plants consumed massive amounts of fuel, making clean water a luxury that cost more than a basic meal. The one resource that had once been the most abundant asset on Earth was becoming more precious than gold.

As Alia walked along the pavement, a voice called out from behind her, "Madam! Madam!"

She stopped and turned around. It was Idris. He had been discharged from the hospital. He was wearing the same soiled clothes having swapped his hospital gown back for that familiar blue uniform. His face was pale, dark circles hollowed his eyes, yet his expression was filled with profound gratitude.

"Madam, I wanted to thank you. You saved my life. I will pray for you until the day I die."

Alia took Idris's hand. "It's alright, Brother Idris. How are you feeling now? What did the doctor say?"

"He prescribed a week of rest, and then light duties. But the company doesn't grant medical leave. I have to report back to the site tomorrow. If I don't, they'll dock my pay, and how will I afford the rent?"

Alia pulled some cash from her purse and pressed it into Idris's palm. At first, he hesitated, drawing his hand back. "No, Madam, you've already done too much." But Alia insisted. She knew this money was merely a temporary narcotic; the root of the problem lay infinitely deeper. The very economic architecture of this city was engineered to survive by exploiting millions of people like Idris. They were functionally invisible, yet without them, not a single skyscraper in this city would stand. They were the ones who built the Burj Khalifa, dredged sand from the ocean floor for Palm Jumeirah, and drove the concrete pillars for the Metro rail. Yet their names were written nowhere. They were merely numbers. And if one perished, that number was simply replaced by another.

Alia resumed her walk. She looked up at the sky, which was choked with grey dust. The sunlight was dull, as if filtered through dirty glass. This city was essentially a vast glass cage inside, a privileged few indulged in luxury, while outside, millions fought simply to survive. Could a city built on such profound inequality endure? History said no. Yet, right up until the moment of collapse, humanity feeds itself on false dreams; Dubai was doing the same.

At home that night, Alia opened her notebook once more. She would record the day's events write about Idris, write about Haji Rashid. In a city where a bottle of water cost five dirhams, she would write its story with absolute tenderness, with meticulous care. Because this story did not belong to Dubai alone it was the narrative of human civilization itself. A civilization that chose to write its destiny not with water, but with oil a destiny that was now turning grey, like a drying river.

Friday, the day of Jumu'ah. Alia woke early and opened her window to find the sky altered. The familiar grey haze that permanently shrouded the city had been replaced by a strange, amber glow as if a colossal dust storm had risen somewhere in the deep desert, casting its long shadow here. The air carried an unnatural, heavy stillness. In a city where some breeze was always stirring, today not a single leaf moved. This silence was terrifying the stillness before a tempest.

After the Jumu'ah prayers, Alia was returning home. Stepping out of the mosque, a wave of blistering air hit her the temperature had risen further, hovering near 50°C. Right then, an emergency alert flashed on her phone: an extremely severe cyclone had formed in the Persian Gulf. The meteorological department had issued an urgent crisis warning. The storm had been named "Harmattan." Named after the dusty desert winds of the Sahara, its trajectory was heading directly toward the United Arab Emirates. Satellite imagery indicated that wind speeds at the core had crossed 160 kilometers per hour and were steadily intensifying

Alia stared at her phone screen. Cyclones were virtually unheard of in this region. The Persian Gulf was geographically sheltered surrounded by landmasses, and its water temperatures had historically never been conducive to cyclogenesis. But last year, a minor cyclone had struck the coast of Oman, serving as a warning shot. Due to rising global temperatures, the warming ocean waters were now generating cyclones in latitudes where they had never existed before. The Persian Gulf had crossed 38°C, providing the storm with immense thermal energy. Scientists had warned a year ago that the Persian Gulf was turning into a cyclone factory, but no one had listened.

The government declared a state of emergency. Orders were issued to evacuate coastal areas. Danger signals were sent to the elite enclaves Palm Jumeirah, JBR, Dubai Marina. The television screens broadcasted continuous loops of the same announcement: "The Dubai Municipality informs all citizens that Extremely Severe Cyclonic Storm Harmattan is projected to make landfall on the UAE coast tomorrow morning.

All residents are urged to move to safe shelters. Emergency services are deployed."

The fundamental problem was that the majority of the population in this city had never experienced a cyclone. They had no idea how to prepare. Basic survival measures taping glass windows, reinforcing doors, stockpiling water and non-perishable food were foreign concepts to many. Furthermore, the city's drainage infrastructure was notoriously fragile. Even a brief downpour caused flash flooding. When three distinct crises heavy rainfall, storm surges, and gale-force winds struck simultaneously, no one knew what the outcome would be; they couldn't even fathom it.

Alia drove straight to her office. An emergency briefing had been called. Arriving in the conference room, she found the tables packed municipality officials, civil defense representatives, high-ranking police officers, and the Institute's scientific team. The tension in the room was palpable. Standing before the screen, Alia analyzed the meteorological models and announced that the probability of Harmattan making landfall was over 90%. Maximum wind speeds could reach 180 kilometers per hour. The coast would be inundated by a massive storm surge, and the rainfall would be so severe that the city's drainage networks would experience total failure.

Hearing this prognosis, the municipality officials sat stunned. A middle-aged Emirati official, dressed in an immaculate white kandura, spoke up, his voice tight: "But Dubai is not engineered for this type of disaster. Our infrastructure was built with the desert in mind. If it rains like this, the entire city will go under water!"

It was a brutal truth. Dubai's roads were designed without adequate stormwater drainage. The drainage pipes were small, few in number, and in many places, choked with desert sand; heavy rain had simply never been a factor in the past. But the climate had shifted, and urban planning had failed to evolve alongside it. Now, the entire city would pay the price for that complacency. The city that believed it had conquered nature was now standing before it, helpless as a child.

Alia couldn't return home that night. She remained in the lab, working alongside her colleagues through the night. The lab was washed in the white glare of generator lights as everyone scrambled some updating data grids, others analyzing satellite frames, some keeping emergency communication channels open. The news broadcasts reported that Harmattan had intensified further as it approached, escalating into a Category 5 storm. Landfall was imminent by morning. Panic buying had begun; supermarket shelves were being stripped bare first the water, then canned goods, followed by batteries, candles, and first-aid supplies. Terror was spreading. Yet beneath this panic lay a strange poetic justice the very nature that humanity had sought to subjugate with an artificial paradise in the desert was now coming to reclaim it. Before nature, human arrogance carries no currency.

Alia had brought her notebook with her. During a quiet interlude, she wrote:

The cyclone will make landfall tomorrow morning. Will this city survive? I do not know. But I know that the terror I am witnessing tonight has played out hundreds of times before in the ledger of history. Babylon, Rome, the Maya all fell. They too believed they were invincible; they too built empires out of stone and declared them eternal. But everything sinks into the ocean of time. Now, it is our turn to drown.

I am not bitter; we had our window of opportunity. We chose to squander it, and now we reap the harvest.

However, my heart breaks for Idris, for Haji Rashid for the legions of laborers who committed no crime, yet will pay the highest price. This is the ultimate tragedy of our age: the architects of ruin escape, while the innocent perish. History has never been an arbiter of justice.

The following morning, Saturday, at exactly 5:40 AM, Harmattan slammed into Dubai. The winds screamed at 170 kilometers per hour so violent that steel street signs were sheared off and launched like missiles, trees were uprooted whole, and the glass windows of skyscrapers shattered, raining shards of glass onto the streets below. Then came the downpour so torrential it felt as if the sky had split open, releasing all its water at once.

Alia couldn't leave the lab; looking down from the upper windows, she watched Sheikh Zayed Road disappear beneath a rising river. Cars floated away, some flipping entirely upside down. People ran for their lives, but the sheer force of the wind made it impossible to stand upright. The power grid failed across the city; mobile networks went down. The city was plunged into total darkness. The howling of the wind sounded like the roar of a monstrous beast. The sky was pitch black, illuminated only by erratic, violent flashes of lightning.

Alia wondered: When will this city see the light again? Will that light be artificial, or will it be the true sun? Or will the lights never come back on at all? Perhaps this is the end of Dubai.

She sat in the darkness, pressing the notebook tight against her chest. Closing her eyes, she thought of her son Hamza when he was a little boy, she used to hold him in her arms, standing by this very window, looking up at the sky. She would say, "Look, baba, the moon has risen." Hamza would ask, "Ammi, how far away is the moon?" Alia would reply, "Very far, my love. But we will reach it one day." Now, it felt as though even the moon had drifted infinitely further away a destination out of reach forever. She remembered her own childhood, the rains of Sylhet, the sound of her mother's voice. Her mother used to say, "Rain is God's mercy." Was this rain a mercy, or a wrath? Alia did not have the answer. She thought that perhaps this city, this entire civilization, was finished. But her letter would endure. Perhaps one day, someone would excavate it from the ruins submerged beneath the sea and read it. And they would know that a group of people had lived here people who fought until the final hour, who tried to preserve hope until the end, who kept writing.

She opened her eyes. Outside, the fury raged unabated. Holding the notebook to her chest, she waited; she waited for the hour when the storm would break, when the light would return, and when the story could begin anew.

The fury of Harmattan finally subsided on the third day. But before it departed, it left an indelible lesson on how hollow human arrogance is when measured against the scale of nature.

Alia had been trapped in the lab for three days. With no electricity and mobile networks down, their sustenance had been limited to a few emergency biscuits and water rations. The generator had eventually sputtered to a halt when its fuel supply ran dry. She had sat in the dark, listening to nothing but the howling wind and the drumming rain. At first, the sound of the rain had been terrifying sounding like stones hurled against the glass. Gradually, that very sound became her sole companion. Finally, on Thursday morning, when she stepped outside, the scene before her eyes made the world seem to stand completely still.

Sheikh Zayed Road a highway that usually carried millions of vehicles daily had been transformed into a sprawling, muddy river. The water was a turbid, choked brown, filled with the wreckage of cars, overturned taxis, shattered billboards, and uprooted date palms. In several sections, the water depth reached six to seven feet. A Toyota Camry floated completely upside down, its wheels pointing toward the sky. The shutters of shops lining both sides of the street had been blasted open, leaving behind thick layers of silt and debris. The air carried a rancid stench a mixture of sewage, oil, burnt plastic, and the unmistakable scent of mortality. The city that had prided itself on being one of the cleanest on Earth had been reduced to a colossal dumpster.

Alia began to walk slowly through the wreckage. The floodwaters had receded slightly, but the streets were still thick with knee-deep mud. Her shoes sank deep into the mire, but she didn't care; her eyes were locked onto the landscape. She looked at the Burj Khalifa it remained standing, certainly, but the entire surrounding plaza looked like a war zone. The lower glass facades had been pulverized, its grand lobby filled with silt and standing water. Palm Jumeirah? Word had reached her that it was partially submerged. The storm surge had swallowed portions of the fronds entirely. The luxury villas that people had purchased for tens of millions of dirhams were now underwater. Luxury, pride, wealth all washed away in a single stroke.

A rescue crew was working nearby. Clad in orange high-visibility jackets and wielding extraction gear, one of them spotted Alia and shouted, "Madam, what are you doing out here? This sector isn't safe. Please evacuate! There are live power lines down in the water, you could be electrocuted!"

Alia stopped. "How many have died?" she asked.

The man fell silent for a moment, then replied softly, "We don't have a count yet. But... the numbers won't be small. The laborer camps took the brunt of it. Entire camps were washed away clean."

Alia asked nothing more. She kept walking. A sudden anxiety struck her she wondered if the notebook she had been writing in last night had been ruined by the moisture. Rushing back up to the lab, she found it resting safely on the table. The room was stifling without air conditioning, but the pages were dry. She breathed a sigh of relief. This notebook was now her only true possession; the singular thing that remained intact amidst the devastation.

She sat down and opened the pages. But before she could begin, her hand trembled. What should she write? A chronicle of the ruin? Or the narrative of human survival hidden beneath the wreckage? She chose the latter. Because history records the chronicles, but human beings write the stories. And it is stories that keep humanity alive. History might eventually forget the exact death toll, the precise date of the landfall, or the peak velocity of the winds. But story would remember the face of Idris, the long sigh of Haji Rashid, the quiet duty of Jamal.

She picked up her pen. Outside, the sun had finally broken through the clouds. The very sunlight that had been a curse a week ago felt like a beacon of hope today the storm had passed, the sky cleared. Yet the temperature was already beginning its relentless ascent 42, 43 degrees.

The city's water supply remained dead. The municipality issued a broadcast: "The situation is under control." Everyone knew it was a lie. But that lie was the only anchor left; the truth was simply too terrifying to face.

Alia sat inside the lab. Around her, colleagues were in motion Rashed was speaking frantically into a receiver, another researcher was updating data on a terminal, and someone was brewing tea in a small electric kettle powered by a briefly revived generator feed. She opened her notebook. The pen's ink had started to dry; she refilled it carefully. Then she began to write slowly, deliberately, shaping each character with profound reverence:

Dear Reader of the Future,

Last night, Cyclone Harmattan finally blew itself out. But the ruin it has left behind will never be erased. You might find this account peculiar a climate scientist witnessing the destruction of a storm and thinking only of stories. I am a scientist, yes, but I am also human. And as a human being, I have seen that even within this devastation, small flowers of hope can bloom. That is what I am recording.

Last night, I was alone in the lab. There was no electricity, only the flicker of a generator lamp for a brief hour. Sitting in that light, I wondered: What did I actually accomplish for this city? I spent my entire life researching, drafting reports, and sounding alarms. But who listened? Who cared? No one.

Then my phone rang. Hamza. He had managed to patch through via a satellite connection. He said, "Ammi, you're alive!" I could hear the immense relief in his voice, mixed with terror. He said, "I'm watching the broadcast; everything is destroyed. Can't you get out of there?" I told him, "I will leave. But not now. I have work to do here."

Hamza grew angry. He said, "Work? What work? That work is the very reason you're trapped there!" I couldn't answer him. Because his words held truth. But there is another truth as well if I flee, who remains? Who will tell the story? Who will explain why this city fell?

Let me tell you the real reason behind the destruction of this city. The cause wasn't merely a cyclone; the cause was that we forgot our limits. We believed that if you poured enough money into the desert, you could manufacture a permanent garden. We believed that if we wrapped ourselves in the cooled air of an AC, we could ignore the burning world outside. We believed petrodollars were a shield against the cosmos. No one stopped to consider that the very ocean from which we pumped our wealth would one day rise to swallow us

The fall of this city is the fall of human hubris. We considered ourselves above nature; today, nature reminded us of our place. She said, "You are nothing. A single storm can erase everything you built."

Yet, when I walked the streets this morning, I saw people digging through the mud to salvage what remained. I saw people transporting the injured to clinics, others distributing bread. None of them were wealthy the majority were laborers like Idris, shopkeepers like Haji Rashid. They are the true soul of this city. The wealthy fled long ago on private jets to Europe and America. But those with nowhere left to flee are the ones standing firm; they are the true owners of this city.

I believe that whatever new city rises from these ruins if one rises at all its foundations will be built by these very people. The ones covered in mud, working on empty stomachs, who still smile as they stand by their fellow man. It is for them that I write this letter. By the time you read this, perhaps that new city will have flourished. And even if it hasn't, let it be known that there was a time when humanity refused to surrender its last shred of hope.

Alia

Finishing the passage, Alia set the pen down. Her hand was shaking the continuous writing had left her fingers aching. Yet her mind felt an uncharacteristic peace, as if, after a long silence, someone had finally heard her voice. She knew no one had listened yet not now but they would one day. And that conviction was what kept her anchored. She closed the notebook and pressed it to her chest. Outside, the sun continued to rise, the heat climbing once more 42, 43, 47 degrees. The water lines remained dry. Yet she found herself thinking that in this very moment, standing amidst the ruins, she was still alive. And to be alive meant there was still work to be done, still words to be spoken, still stories left untold.

Alia stepped outside the office building. The distant thrum of helicopters echoed overhead the rescue operations were underway. International aid had been requested. Yet she knew that while aid would arrive, it could never restore the city to what it was; for that old reality was the very trap that had led them here. From this point forward, humanity had to reinvent its path. And the beginning of that new path started right here with a notebook, a pen, and the chronicles of ordinary lives.

By the fourth week following the storm, the city began to slowly open its eyes. Electricity had been restored to most sectors, though blackouts remained a structured reality the grid went dark for at least six hours every day. The water supply was beginning to stabilize, though the pressure was incredibly low, and residents were instructed to boil every drop to prevent the spread of waterborne outbreaks. The mud had been cleared from the thoroughfares by bulldozers, though drifts of desert sand and shards of shattered glass still accumulated in the corners. Temporary relief camps had been erected across the city, housing thousands of displaced laborers whose camps had been obliterated, who had lost their livelihoods, or whose companies had simply vanished overnight.

Every morning before heading to the laboratory, Alia made a point to visit the site of the 'Green Roots' initiative. The abandoned plot of land that had once been a dumping ground was now lined with rows of vegetable seedlings tomatoes, lettuce, peppers, eggplants, cucumbers; life pulsing in every leaf. Idris had been appointed the head supervisor of the project. Waking at six every morning, his first task was to tend to the seedlings using a water-saving drip irrigation system he had mastered at the training center. He carefully inspected the plants, checking for pests or root rot. Alia watched him in wonder the man who, just a month prior, had been on the verge of death from heatstroke had now found a completely renewed lease on life.

"How are you finding it, Brother Idris?"

Idris, who had been kneeling on the earth inspecting the tip of a tomato vine, looked up and smiled. "Madam, I could never forget the smell of the earth. As a boy in Comilla, I used to work the fields with my father. The scent of the soil, the rice, the jute it's woven into my blood. Spending these past four years locked inside concrete, I had forgotten what it felt like to live as a human being. This garden has made me human again."

Tears welled in Alia's eyes. Listening to Idris, she was reminded of her mother. Her mother used to say, "There is nothing greater than the earth. The earth is everything. We come from the soil, and to the soil we shall return." How absolute that truth was.

The Green Roots project had now drawn nearly thirty volunteers. The majority were migrant workers from Bangladesh, Pakistan, India, Nepal, and the Philippines. Every one of them had been broken by the storm in some way some had lost their employment, others their shelter, and many were the sole providers for families back home. Yet within this garden, they were rediscovering a sense of purpose. They weren't merely cultivating vegetables; they were acquiring skills in solar panel maintenance, water purification, basic electrical repair, and composting. Armed with these skills, they were beginning to find employment elsewhere in the city. Some had started rooftop gardens on their apartment blocks; others had launched small-scale nurseries. The project had outgrown its definition; it had transformed into a movement.

At noon, Alia returned to the lab. Rashed was waiting for her, hunched over a new set of data models. "Dr. Alia, look at these satellite captures from today. The water temperature in the Persian Gulf hasn't dropped; in fact, it's climbing. The models suggest a high probability of another cyclone forming next month."

Alia stared at the monitor, a deep anxiety settling over her. If another cyclone struck, nothing would remain of this city. The wounds of the first storm had not yet closed; a second strike would be fatal. She asked, "Has the ministry been briefed?"

Rashed shrugged, his expression weary. "They've been notified. But they are so consumed with the immediate reconstruction of the commercial sectors that the warnings are being dismissed. Their consensus is: 'We just survived a historic storm, why would another one happen so soon?'"

Alia let out a long sigh. This complacency was their greatest adversary. Every time a crisis passed, humanity assumed the danger was gone forever. Yet the danger returned, and each time, it returned on a grander scale.

Later that afternoon, Alia visited the clinic. The cough that had started the previous week refused to clear. She couldn't sleep at night her breathing was shallow, and a heavy pressure sat on her chest. Her chronic respiratory condition was deteriorating. Dr. Farhan, a middle-aged pulmonologist of Syrian descent, looked deeply concerned as he reviewed her CT scans and blood panels.

"Dr. Alia, the pulmonary fibrosis has advanced significantly. The degeneration over the last six months has been aggressive. The air in this city has become toxic to your system. I told you before, and I must emphasize it now if you want to survive, you must leave this city. Your son is in London; go there and seek treatment."

Alia remained silent for a long moment. Then, she asked softly, "Doctor, if I leave, what happens to the people here? My research, my team, Green Rots am I to simply abandon everything?"

"You can only advocate for others if you are alive to do so. If you succumb to this illness, you can do nothing for them."

Alia knew the doctor's counsel was entirely rational. Yet she also recognized that if she fled simply to preserve her own life, leaving everyone else behind, that survival would feel hollow. The struggle she had entered was not a personal one it was the collective struggle of humanity. And as someone who had helped lead that struggle, she could not abandon the field.

Leaving the clinic, Alia found herself walking back toward Deira. Her spirit felt heavy; the doctor's warnings echoed relentlessly in her mind. The thought of her own mortality did not terrify her death was an inevitability for everyone. But she worried about the work left unfinished. Would she have the time to complete her letter? Would Green Roots endure? Would Idris be able to secure the future he envisioned for his daughters?

Her steps brought her back to Haji Rashid's eatery. Half of the structure remained unrepaired the storm had ripped away the roof, which was now replaced by a heavy tarp. Yet Haji Rashid continued to cook. Seeing Alia, his face lit up with a brilliant smile.

"Welcome, Madam, welcome. I've prepared something special today mutton nihari. A true Pakistani specialty. Tell me, how is your health?"

Alia took a seat. "Not good, Haji Sahab. The doctor advises me to leave the city. But my heart refuses to let go."

Haji Rashid went quiet for a moment, wiping down the table. Then he spoke softly, "Madam, let me share something with you. I arrived in this city thirty years ago. Back then, there were only a handful of concrete structures in Deira. I watched this city rise with my own eyes, and now I've watched it break. But I know we will build it again. Because the ground that holds human love can never be permanently destroyed. The work you are doing here is the highest expression of love for this city. If you leave, that love diminishes. But if you stay, others will see you and learn how to love this place too."

Alia wiped a tear from her cheek. What profound wisdom and deep tenderness lived in the words of this ordinary man. It was for people like him that she wanted to live; it was for them she would continue to fight.

Returning home that evening, she opened her notebook and recorded the day's reflection:

Today, the physician urged me to abandon this city. But Haji Rashid reminded me exactly why I remain it is the pull of love. My affection for this city, for its people, for this very soil is what sustains me. I know my time is limited, but for as long as I draw breath, I want to scatter this love. Because love is the singular force that outlasts destruction.

By mid-August, the international press corps began arriving in numbers. Crews from the BBC, CNN, Al Jazeera, and Reuters arrived to document the aftermath of Harmattan. However, their coverage focused primarily on the destruction within the wealthy enclaves the ruined villas of Palm Jumeirah, the flooded parking basements of Dubai Mall, the fractured windows of the Burj Khalifa. The wreckage of the laborer camps drew little interest from their cameras.

Alia recognized an opportunity that could not be squandered. Collaborating with Rashed and a few colleagues, she produced an independent documentary short titled: Dubai: The Unseen Disaster. In it, they laid bare how the migrant worker population had borne the catastrophic weight of Harmattan. They documented the obliterated camps, the families left homeless, the lack of medical access, and the individuals facing starvation after being abandoned by their employers. Idris was featured prominently speaking candidly about his ordeal, his near-fatal heatstroke, the systemic injustices of the contracting system, and the renewal he had found through Green Roots.

The production took nearly three weeks to finalize. Once completed, Alia dispatched the project to international media networks. Initially, it was met with silence. But when a journalist from Al Jazeera viewed the footage, struck by its urgency, he traveled down to Naif to inspect the conditions of the laborers firsthand. The story went viral. The hashtag

#DubaiUnseen began trending across global platforms. Voices from every corner of the world began demanding accountability.

Yet this visibility carried consequences. Certain factions within the government were furious. They viewed Alia's work as a deliberate campaign to tarnish the city's international prestige. A prominent developer called her terminal directly, his voice dropping all pretense of civility: "Dr. Alia, cease these activities immediately. If you persist, the institutional funding for your research will be terminated."

Alia refused to be intimidated; the threat only hardened her resolve. She understood that the truth she held was so potent that the powerful were desperate to suppress it. And when power seeks to silence a truth, it means revealing it has become an absolute necessity.

Late one evening, she called Hamza. For the first time in years, she did not speak of their estrangement, but said softly, "Hamza, do you remember what your father told me long ago? He said, 'Whatever you write, write it with this pen. This is your voice.' I am writing with that very pen, baba. And now, the world is finally beginning to hear my voice."

Hamza was silent for a continuous moment before responding, his voice thick with emotion: "Ammi, I am immensely proud of you. But I also need you to stay alive."

"I will live on, my son. Perhaps not in this physical form, but in my words, in my work, and in the love I leave behind."

Hamza began to weep silently on the other end of the line. Alia couldn't restrain her own tears. The invisible distance that had strained the relationship between mother and son for so long dissolved in that quiet moment.

By the arrival of September, Alia's physical condition deteriorated further. Dr. Farhan informed her that the fibrosis had advanced to a stage where managing without continuous oxygen supplementation was no longer viable. She began using a portable oxygen concentrator a compact device she carried over her shoulder, supplying oxygen via a nasal cannula. When she entered the laboratory carrying the machine, her colleagues looked away in sorrow, but Alia offered a lighthearted smile.

"Don't look so grim, think of it as my latest fashion accessory," she joked.

But everyone understood the agony masked behind the humor. Rashed wiped his eyes, pleading, "Dr. Alia, please work from home. We will bring the daily updates and data sheets to your apartment."

"No, Rashed. If I sit isolated at home, my mind will give up. This work is what keeps me alive. Without it, my survival has no purpose."

During this period, the Green Roots initiative expanded rapidly. The original plot in Deira could no longer contain the project; three new branches were established one in Sonapur, one in Al Qusais, and another near the industrial worker barracks in Jebel Ali. Each location replicated the original blueprint: rooftop cultivation, vertical farming modules, rainwater harvesting systems, solar arrays, and a training center. The initiative now encompassed nearly two hundred active participants; individuals who had arrived with no prior agricultural knowledge were now serving as instructors themselves.

Idris had evolved beyond his role as a supervisor; he had become an educator. Every week, he trained new influxes of volunteers demonstrating how to prepare the soil beds, configure the drip lines, and manage organic composting. His voice held a quiet authority, his eyes bright with a new vision. Watching him from a distance, Alia thought to herself: This is what fulfillment looks like.

One morning, Idris approached her, his face radiant. "Madam, I have news. My daughters are coming. Nadia and Nishat are traveling here. I've saved enough money to enroll them in school; I am going to ensure they get their education."

Alia was overcome with joy. "Truly? How beautiful, Idris! When do they arrive?"

"Next month. My wife was meant to come as well, but she decided to remain in the village to care for my aging mother for a bit longer; she will join us later."

This brief piece of news filled Alia with a profound sense of peace. She realized that change was manifesting slowly, in small increments, but shifting nonetheless. She had personally intervened to ensure Idris's daughters secured placement in a school. In Dubai, enrolling the children of migrant laborers into the educational system was an uphill battle fraught with bureaucratic red tape, high fees, and endless systemic hurdles. But Alia had utilized her personal network to secure their entry into a private academy. The headmistress had initially been hesitant, but Alia had looked her in the eye and asked: "This girl's father built this city with his bare hands. If his child is denied an education here, what value does this city actually possess?" The headmistress had signed the papers.

Alia realized that these quiet victories were the ones that truly mattered. Transforming the trajectory of a single human life was far more significant than drafting abstract policy whitepapers. She recorded in her journal:

Today, Idris told me his daughters are arriving. My heart danced with pure joy. This single piece of news validates every ounce of exhaustion, every struggle I've endured. I see now that change never descends from the heavens in a grand gesture; it occurs quietly within the human heart, within an individual life, within a home. When one family transforms, the collective community begins to shift. I want to dissolve into that current of transformation like a single drop of water. To lose myself in it completely, without a shred of regret.

In the second week of October, Alia was formally admitted to the hospital. Her respiratory distress had escalated to a degree that home management was no longer viable. Dr. Farhan stated, "Your lungs can no longer oxygenate your blood effectively on their own. We must transition you to full institutional support and run a series of advanced diagnostics."

Alia lay in her hospital bed. Through the glass window, she could see the sky outside today it was a pristine, brilliant blue, showing no scars of the tempest that had battered it. Yet Alia knew that beneath this serene sky lay the potential for countless future storms. Like her own failing anatomy, nature was growing increasingly volatile.

Even from her bed, she refused to abandon her writing. The notebook and her pen remained on the tray beside her. Whenever a brief surge of strength returned, she would write. The nursing staff had initially tried to dissuade her "You must rest, Madam, there is no need for this exertion." But eventually, they ceased their interventions. They recognized that this writing was Alia's primary life support.

One afternoon, Hamza walked into her room. Alia looked up in astonishment; he was supposed to be in London. "When did you arrive, baba?"

Hamza rushed forward, taking her frail hand in his. "I've taken an extended leave, Ammi. I am staying right here with you."

Alia could not stem the tears that flowed down her face. Hamza's presence was a greater medicine than any drug the clinic could provide. Mother and son sat together for hours, speaking softly of things long left unsaid. Alia told him, "I've left a few things for you in the cabinet at the apartment. The notebook, the pen, and some of my texts. Ensure you take them."

Hamza began to speak but stopped, his voice catching in his throat. After a quiet moment, he asked, "Ammi, are you still writing?"

"I am. I haven't finished it; perhaps I won't be granted the time to finish it. But I am writing as much as my hand allows."

"I will finish it, Ammi. Whatever you are unable to leave on the page, I will complete it for you."

Alia gazed into her son's face. In his eyes, she recognized the same fire that had once defined her own youth. That fire passes from one generation to the next; that is the true continuity of civilization not merely genes or material culture, but the lineage of dreams and resistance.

That night, after Hamza left the ward, Alia wrote:

Hamza came today my son, my pride. I saw in his eyes the exact fire that once burned within me. I know that when my physical form fails, that light will not go out. Hamza will preserve it, Idris will carry it forward, Rashed will keep it alive. Every human being who loves this world enough to fight for it will keep that fire burning. I am merely a single spark; it is only natural that a wildfire should rise from it. My light may dim, but that is of no consequence as long as the fire continues to burn.

By the beginning of November, Alia's condition took a terminal turn. Alongside her respiratory failure, her renal systems began to shut down. Dr. Farhan initiated dialysis treatments, though he knew it was a holding action; her body was losing its final battle.

Yet Alia's spirit remained unyielding. Every day, she committed something to the page sometimes an entire leaf, sometimes two lines, occasionally just a singular fragment of a sentence. But she never stopped. Her writing had transitioned beyond the definition of a letter it had become a manifesto, a historical testament, a living witness; the final signature left for the future.

One morning, Idris arrived at the clinic accompanied by his two daughters. Nadia and Nishat both possessing their father's deep complexion and bright, inquisitive eyes. They initially hesitated upon entering the room, intimidated by the array of medical machinery surrounding the frail woman in the bed. But when Alia offered a warm smile and gestured for them to approach, they stepped forward.

"Nadia, Nishat! Look how tall you both have grown! How are you liking Dubai?"

Nadia smiled shyly. "I love it, Auntie. I've made many friends at school already."

Nishat looked up, her voice clear: "I want to become a doctor when I grow up, just like you."

Alia's eyes welled with tears. Nishat wanted to practice medicine because of her. This is legacy, she thought. Taking Nishat's small hand, she whispered, "You absolutely will, my dear. You will become a doctor, and you will be a far greater one than I ever was."

Idris stood by the edge of the bed, weeping silently. He knew this woman had not only preserved his life but had shown him how to rebuild it from the ashes. Now, he was watching her slowly slip away. What could he do? How could he ever repay such a debt?

Alia, reading the anguish on his face, spoke softly: "Brother Idris, do not carry the weight of any debt. Your only duty is to raise these girls make them strong, make them good human beings. They are the ones who will alter the course of this world. When I am gone, you must remain continue to love this city, continue to love this earth."

Idris nodded, unable to summon words through his grief, but his tears spoke for him.

That evening, Alia opened the notebook for the final time. Her hand shook violently as she held the pen. She wrote with immense difficulty, forcing the words onto the paper:

Dear Future,

I can feel the twilight arriving. My anatomy tells me it can endure no further, though my spirit insists there is still so much left to say. I must accept the verdict of the body over the spirit. Yet I know with absolute certainty that the expiration of the flesh does not mean the expiration of the soul. The soul endures in the written word, in the work left behind, in the collective memory of those we touched.

What I have accomplished is incredibly modest a small garden plot, a handful of seedlings, the training of a few individuals. This alone will not alter the global climate. But it is my profound belief that from this tiny garden, a vast forest will one day rise; for every great tree on this Earth was once nothing more than a tiny seed. I have planted the seed; your task is to nurture it into a sanctuary.

My final counsel to you learn to love. Learn to love nature, learn to love humanity. Because love is the only entity that outlasts destruction. Hatred consumes, greed decimates, arrogance tears down; but love preserves, constructs, and sustains.I am departing, but you are here. I leave you all my love. This affection is my final letter to you.

Alia

Alia Al-Qasemi passed away in the quiet pre-dawn hours of November 23, 2075. Her final directive had been explicit: no artificial life support was to be maintained. "When nature calls me back, let me go. Do not chain me to a machine," she had told Dr. Farhan. The clinical record noted the cause of death as total respiratory failure. But those who knew her understood that she had not truly ended; she had simply paused taking a necessary rest at the conclusion of a long journey.

The news traveled swiftly. First through the corridors of the Institute, then among the volunteers of the Green Roots initiative, before spreading across global networks. The hashtag #DrAliaAlQasemi began to trend worldwide. The woman who had lived in relative academic obscurity just months prior had been transformed into an international symbol an icon of resilience for the planet, of speaking truth until the final breath, and of unconditional love.

Hamza remained in Dubai, having quietly canceled his return ticket to London. He resolved to complete his mother's unfinished work. Standing in her apartment, he found the notebook resting on her desk, accompanied by a brief note: "For Hamza." He opened the pages and began to read. He read through the night, discovering his own childhood woven into her handwriting, encountering her love, and inheriting her vision. By the time he reached the final entry, the sun was rising over the horizon a deep, crimson sun filtered through the desert dust, herald of a new day.

The funeral service was held at the Al-Farooq Mosque. The congregation was vast her scientific colleagues, the green volunteers, Haji Rashid, Idris standing proudly with his two daughters, Mrs. Rogers, and even several government officials who had once vehemently opposed her work. Death possesses a quiet power to dissolve hostility; or perhaps, an unuttered guilt moved them. The woman whose warnings they had dismissed was gone, and now, amidst the changing climate, they were beginning to realize how terribly accurate she had been.

Following the service, Idris approached Hamza. Taking his hand, he said, "Brother, what your mother planted here, we will never let it die. This city will never forget her. I have made my daughters promise that they will spend their lives protecting this Earth."

Hamza squeezed Idris's hand. "My only wish, Brother Idris, is that your daughters fulfill the dream my mother held."

A few weeks later, while organizing his mother's desk drawers, Hamza discovered a collection of documents meticulous schematics for the expansion of Green Roots, proposals for new urban farming nodes, and blueprints for international adaptation frameworks. Even while confined to her clinic bed, she had been engineering this expansive vision; as if she knew her own time was short, but her dream had to outlive her. Hamza realized that executing these designs would be the ultimate monument to her memory.

Over the succeeding months, Hamza dedicated himself to advancing her work. He was not an environmental scientist by training he had studied macroeconomics and had been employed at an international investment bank. Yet his mother's legacy pulled him into an entirely different universe. He resigned from his banking position, committing his entire focus to the Green Roots Foundation.

The initial transition was incredibly demanding. The divergence between commercial enterprise and grassroots community work was vast. Hamza made numerous missteps, encountering moments of deep frustration; but every time he faltered, he recalled his mother's mantra:

"The struggle itself is what defines a life." And so, he fought on. Gradually, he mastered the art of scaling a grassroots community project into a permanent, self-sustaining social enterprise.

Green Roots was no longer contained within Dubai; its networks expanded into Abu Dhabi, Sharjah, and Ajman. The methodology proved so profoundly effective that the United Nations Sustainable Development Division initiated contact, seeking to deploy the Green Roots framework in other climate-stressed urban centers globally. Hamza agreed, but inserted a non-negotiable clause: "This initiative shall never fall under corporate or state ownership. It must remain an initiative of the people, managed by the people, for the people."

Concurrently, Hamza began working on the manuscript of his mother's notebook. He personally translated her entries into Arabic, Urdu, Hindi, and Bengali. He wanted her voice to reach every corner of the world, allowing humanity to share in her vision.

He titled the volume: The Last Letter of the Century: From Dubai with Love. Independent publishers were initially hesitant; one acquisition editor remarked, "The public is exhausted by climate literature; it's simply too bleak a subject." But Hamza was unyielding. Eventually, a small, independent press agreed to print a modest run.

Upon its release, the book sold a mere seven hundred copies in its inaugural week. But then, through word of mouth, its readership grew exponentially. It resonated profoundly with the younger generation, who recognized the trajectory of their own futures within Alia's reflections. They began to organize, establishing the "Alia Foundation" an organization dedicated to advocating sustainable urban living, expanding community agriculture, and intensifying public awareness around climate thresholds.

Hamza watched as his mother achieved a quiet immortality. Her words, her dream, and her love had taken root in the hearts of millions.

To mark the fifth anniversary of the Green Roots project, a gathering was organized at the site of their very first harvest in Deira the plot where

Alia and Idris had sowed the first seeds together. The once-abandoned lot had been transformed into a lush, vibrant public sanctuary, officially named "The Alia Garden." It was surrounded by green canopies, floral borders, and vegetable patches; at its center sat a modest pavilion housing a public environmental library.

People had traveled from across the globe to attend the anniversary scientists, activists, students, agriculturalists, and laborers. A singular question united them: How do we carry Alia's vision into the next century?

Hamza delivered the opening address. "My mother was neither a celebrity nor a public icon," he spoke into the microphone. "She was an ordinary individual who resolved to do something extraordinary. Whether she succeeded is for the future to decide. But she fought for that vision until her very last breath. That effort is what matters. If each of us, within our own spheres, commits to a small measure of effort, the trajectory of this world can shift. And even if it doesn't, we must make the attempt. Because to refuse the struggle is to succumb to a premature spiritual death. To live fully is to persist in the effort."

Idris took the podium next. His voice retained the thick accent of his home district in Comilla, but his delivery carried an absolute, grounded confidence. "There was a day when I lay on the asphalt of this city, waiting to die. My head was spinning from the heat, my body was burning, and the world looked away. Only one human being stopped to save me my Alia Madam. She didn't just preserve my life; she taught me how to live it with dignity. Today, I live, and I teach others how to survive. That is her legacy. I ask you all to carry that lesson home survive, and help others survive."

At the conclusion of the ceremony, Hamza announced the establishment of a new global initiative funded by the foundation: "The Alia Al-Qasemi Scholarship for Climate Action." The fellowship would be awarded annually to five young innovators working on grassroots climate resilience within their vulnerable home communities.

Nishat, Idris's younger daughter, now a bright secondary school student, stood up from the audience and spoke clearly: "I will be applying for this fellowship. I am going to study medicine, and I will return to provide healthcare to the people of my father's village without charge. I am going to fulfill the vision that Alia Auntie gave me."

Hamza watched her, deeply moved. His mother's dream had truly taken flight, soaring beyond her life.

As the event concluded, twilight settled over the Alia Garden. The crowds dissipated, leaving only Hamza and Idris sitting together on a wooden bench beneath the arbor. The stars were beginning to appear singly at first, then multiplying into a dense canopy. The night sky of Dubai rarely revealed its stars, usually obliterated by the intense artificial glare of the metropolis. But here, within this quiet, green sanctuary, nature seemed to unveil itself completely.

Idris broke the silence. "Brother, your mother used to say that the world as we knew it would end. But I look at the seeds she left behind, and I wonder can a seed like that ever truly end? A tree may wither, but its fruit drops the seeds for a new forest. Within this garden, she remains alive. Every morning when I come here to tend the vines, I feel her presence."

Hamza nodded, looking up at the leaves. "You know, Brother Idris, on the night before she passed, my mother wrote: 'Love is the singular force that outlasts destruction.' I look around us now, and I realize this garden is the physical manifestation of that love. It is woven into the root system of every plant here. That love is immortal."

Idris remained quiet for a long moment, then said softly, "My daughter Nishat told me she intends to write a biography of Alia Madam. I told her write it, but write it with absolute truth, so that the world understands exactly what this woman gave to us."

Hamza smiled. "Let her write it. My hope is that gardens like this one rise in every city across the globe. Even the desert can bloom, if humanity remembers how to love the earth."

The night deepened. The two men rose from the bench, walking toward the garden gates. Near the exit, Idris paused, turning back to look at the dark, rustling canopy of the sanctuary one last time. In a voice thick with emotion, he whispered, "May the Almighty grant her peace in the highest heavens. She was truly a blessing to us."

Hamza said nothing, but placed a steady, reassuring hand on Idris's shoulder. Then, the two men stepped out into the night together an Emirati man and a Bangladeshi worker. Divided by nationality, by language, and by background yet bound by an absolute, shared devotion. A devotion that held them in an unbreakable bond.

In the distance, the artificial lights of the Burj Khalifa flared against the night sky as the metropolis returned to its familiar, glittering choreography. But if one looked closely, with a perspective tempered by tenderness, they would see beneath that artificial glare, right down at the level of the earth, small green shoots pushing through the soil. Shoots that would one day mature into massive trees, under whose shade weary travelers would find shelter, and whose blossoms would scent the hot desert air. On that day, the passersby might not know the name of the woman who had planted them. But that knowledge was unnecessary; the existence of the tree was all that mattered.

And that enduring, green sanctuary bore the name Alia.

THE END

ABOUT THE AUTHOR

Rajuan Ahmed writes quietly. His stories don't shout. They build from small moments, the kind most people overlook. Ordinary lives supply his material; he turns their everyday struggles into lasting fiction.

He started in the early nineties. His first short story, "Abhagi" (The Unfortunate Woman), was written in 1991, during his college years. During university, he turned to the visual medium. Screenwriting pulled him in. "Marriage Station," his debut television drama, took off unexpectedly and made his name in television. More dramas followed across multiple channels, and audiences embraced them.

He finished his Master's in 1999 and joined a corporation. The day job did not kill his writing. Through those years, he continued publishing short stories, essays, and columns in national dailies, earning a steady readership.

He also writes songs. His lyrics carry a distinct rhythm, and they have found a wide audience. National awards have recognized his contributions to both literature and music.

His range spans short stories, novels, poetry, television dramas, songs, and essays. He works in all of them with a distinct voice, understated but sharp, leaning on small psychological turns rather than grand declarations. His notable books include Ek Mutho Golpo, Sondeher Chhaya, Mayabi Muhurto, Shonkher Shopoth, Shobde Tumi, Alokkchhaya, Shopner Chakri, Shesh Surjer Alo, The Days of Gaza, Love of the Forest Bird, Death Break, The Shadow of War, Opekkhhar Chhaya, and O-Jol Nirbashon.

He avoids publicity. Art, he believes, holds its own weight. So he writes in silence, tracking what people leave unsaid and how time presses on the human mind.

Don't miss out!

Visit the website below and you can sign up to receive emails whenever Rajuan Ahmed publishes a new book. There's no charge and no obligation.

https://books2read.com/r/B-A-MJSSE-RDIOJ

BOOKS 2 READ

Connecting independent readers to independent writers.

Did you love *Dubai 2075*? Then you should read *Beyond the Naf*[1] by Rajuan Ahmed!

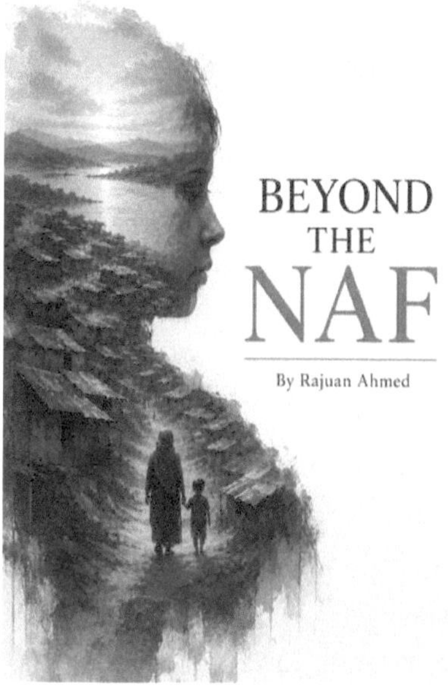

[2]

In 1799, when East India Company scholar Francis Buchanan first recorded the name "Rooinga" along the misty estuaries of Arakan, he documented an indigenous identity deeply rooted in the fertile soil of the Naf River valley. Spanning over two centuries, *Beyond the Naf* by Rajuan Ahmed is an epic, multi-generational novel of identity, belonging, cultural heritage, and tragic displacement.

At the heart of the story is the family of Hamid Miah, a devoted farmer who builds his traditional mud house (*maitta ghor*) and flourishing betel orchards in the village of Shalbagan. Bound by ancestral honor, Hamid instills a deep connection to their motherland in his son, Ismail, who grows up to become a respected primary school teacher.

---

1. https://books2read.com/u/m2gg5d
2. https://books2read.com/u/m2gg5d

Together, they clear the alluvial plains, cultivate golden rice fields, and erect a proud wooden *tottar ghor*, standing resilient through changing political eras.

Yet as Burma transitions from British colonial rule to post-war independence, the promise of equal rights quickly unravels. The 1962 military coup triggers decades of systematic exclusion, land confiscation, and state-sponsored violence. When the infamous 1982 Citizenship Law strips the Rohingya of their fundamental rights, it renders millions stateless overnight in the land of their forefathers. Through repeated military clearance operations, Ismail and his son Zahir struggle to defend their home, guided by the sacred final words of their patriarch: *"Never leave this soil."*

As brutal violence engulfs Arakan once more in August 2017, the family faces its ultimate catastrophe. With burning villages reddening the night sky and gunfire echoing along the riverbanks, Zahir must navigate a heart-wrenching choice between his sacred ancestral roots and his family's sheer survival.

Rich in historical detail, traditional craftsmanship, and human warmth, *Beyond the Naf* breathes vivid life into a community before it was uprooted. Rajuan Ahmed delivers an unforgettable literary monument to the Rohingya people a poignant narrative of love, dignity, and endurance.

Also by Rajuan Ahmed

01

Beyond the Naf
Dubai 2075
Exiled Dreams
Quantum Wizards
In the Land of Mirages
Love of the Forest Bird
The Days of Gaza
Geneva Camp
We Are All Mad
The Shadow of War

Standalone

Death Break

About the Author

Rajuan Ahmed writes quietly. His stories don't shout. They build from small moments, the kind most people overlook. Ordinary lives supply his material; he turns their everyday struggles into lasting fiction.

He started in the early nineties. His first short story, "Abhagi" (The Unfortunate Woman), was written in 1991, during his college years. During university, he turned to the visual medium. Screenwriting pulled him in. "Marriage Station," his debut television drama, took off unexpectedly and made his name in television. More dramas followed across multiple channels, and audiences embraced them.

He finished his Master's in 1999 and joined a corporation. The day job did not kill his writing. Through those years, he continued publishing short stories, essays, and columns in national dailies, earning a steady readership.

He also writes songs. His lyrics carry a distinct rhythm, and they have found a wide audience. National awards have recognized his contributions to both literature and music.

His range spans short stories, novels, poetry, television dramas, songs, and essays. He works in all of them with a distinct voice, understated but sharp, leaning on small psychological turns rather than grand declarations. His notable books include Ek Mutho Golpo, Sondeher Chhaya, Mayabi Muhurto, Shonkher Shopoth, Shobde Tumi, Alokkchhaya, Shopner Chakri, Shesh Surjer Alo, The Days of Gaza, Love of the Forest Bird, Death Break, The Shadow of War, Opekkhhar Chhaya, and O-Jol Nirbashon.

He avoids publicity. Art, he believes, holds its own weight. So he writes in silence, tracking what people leave unsaid and how time presses on the human mind.

About the Publisher

Headroom Publishers is dedicated to discovering bold voices and publishing books that inspire, inform, and endure. Committed to quality editing, creative design, and global distribution, we help authors reach readers worldwide while celebrating ideas, imagination, and creating a lasting impact.

www.ingramcontent.com/pod-product-compliance
Lightning Source LLC
LaVergne TN
LVHW100201061126
851857LV00010B/348